Abdelhafid Mimouni

Bioinorganics decipher Clostridium botulinum

Abdelhafid Mimouni

Bioinorganics decipher Clostridium botulinum

ScienciaScripts

Imprint

Any brand names and product names mentioned in this book are subject to trademark, brand or patent protection and are trademarks or registered trademarks of their respective holders. The use of brand names, product names, common names, trade names, product descriptions etc. even without a particular marking in this work is in no way to be construed to mean that such names may be regarded as unrestricted in respect of trademark and brand protection legislation and could thus be used by anyone.

Cover image: www.ingimage.com

This book is a translation from the original published under ISBN 978-620-6-71437-8.

Publisher:
Sciencia Scripts
is a trademark of
Dodo Books Indian Ocean Ltd. and OmniScriptum S.R.L publishing group

120 High Road, East Finchley, London, N2 9ED, United Kingdom
Str. Armeneasca 28/1, office 1, Chisinau MD-2012, Republic of Moldova, Europe
Printed at: see last page
ISBN: 978-620-8-04100-7

Bioinorganics decipher Clostridium botulinum

Author: Dr Abdelhafid Mimouni is an independent researcher specialising in bioinorganic systems chemistry, with extensive expertise in macromolecular synthesis and characterisation. He obtained his PhD in chemistry from the University of Paris XII in 1997 and a Diplôme d'Études Approfondies in bioinorganic systems from the University of Paris XI in 1993.

Summary:

This in-depth book on Clostridium botulinum and its neurotoxin, botulinum toxin, explores in detail the biological and chemical aspects of this anaerobic bacterium. Focusing on bioinorganic chemistry, it examines how Clostridium botulinum uses metals and metalloenzymes to metabolise and protect itself in hostile environments. The structure and mechanism of action of botulinum toxin are explained in detail, revealing how it interferes with nerve transmission for innovative applications. Ethical considerations and regulatory challenges are addressed, as well as recent technological advances that improve its efficacy and safety. In conclusion, this book offers an in-depth, interdisciplinary exploration of Clostridium botulinum and botulinum toxin, highlighting their continuing potential for scientific research and technological innovation.

Table of Contents

Introduction

Overview of Clostridium botulinum

Clostridium botulinum is an anaerobic, Gram-positive, spore-forming bacterium, widely known for its ability to produce one of the world's most potent toxins: botulinum toxin. This bacterium occurs naturally in soil, water and sediment, and can survive in harsh environmental conditions thanks to its ability to form resistant spores. When it finds a suitable environment, particularly in the absence of oxygen, Clostridium botulinum can develop and produce toxins that inhibit the release of acetylcholine, a molecule essential for transmitting nerve signals to muscles.

History and discovery of botulinum toxin

The story of Clostridium botulinum and its toxin begins in the early 19th century. In 1793, cases of food poisoning in the Württemberg region of Germany attracted the attention of the physician Justinus Kerner. Between 1817 and 1822, Kerner published several works describing the symptoms of a mysterious illness he called "poisoned salmon", which he attributed to a toxin present in contaminated sausages. Kerner suspected that this toxin acted on the nervous system and even proposed its therapeutic use to treat certain neurological disorders.

The term "botulism" is derived from the Latin word "botulus" meaning "sausage", in reference to Kerner's initial observations. However, it was not until the end of the 19th century that the link between the bacterium and the toxin was formally established. In 1895, a Belgian epidemiologist, Emile van Ermengem, identified Clostridium botulinum as the causative agent of food poisoning following the consumption of contaminated ham. Van Ermengem succeeded in isolating the bacterium and demonstrating that the toxin it produced was responsible for the symptoms observed in the patients.

Over the decades, research into Clostridium botulinum and botulinum toxin has progressed significantly. During the Second World War, scientists began to explore the potential use of the toxin as a biological weapon because of its extreme potency. However, this research also led to crucial discoveries about the therapeutic use of botulinum toxin. In the 1970s, Dr Alan Scott used botulinum toxin type A to treat strabismus, marking the beginning of the medical use of this substance.

Today, botulinum toxin is used in a variety of medical and cosmetic fields. It is commonly used to treat conditions such as muscle spasms, cervical dystonia, hyperhidrosis, and even to reduce facial wrinkles. Botulinum toxin treatments require precise, controlled administration by qualified healthcare professionals.

In terms of treatment, several antibiotics are often used to manage Clostridium botulinum infections. Penicillins and metronidazoles are commonly prescribed to eradicate the bacterial infection, although the primary management of botulism cases often involves intensive supportive care to counter the effects of the toxin.

This introduction lays the foundations for a more in-depth exploration of the biochemistry of Clostridium botulinum, the mechanisms of action of botulinum toxin, and its clinical applications and implications, while providing a historical overview crucial to understanding the evolution of our knowledge and use of this bacterium and its toxin.

Chapter 1: Biology of Clostridium botulinum

Morphology and Classification

Clostridium botulinum is a bacterium belonging to the Clostridiaceae family. Morphologically, it takes the form of straight or slightly curved bacilli, typically 4 to 6 micrometres long and around 1 micrometre wide. It is Gram-positive, which means that it retains the crystal violet dye in the Gram stain, due to the particular structure of its peptidoglycan-rich cell wall.

Classified as a strict anaerobic bacterium, C. botulinum can only survive in the absence of oxygen. This characteristic stems from its enzymatic structure and metabolism, which do not tolerate the presence of oxygen. However, phylogenetic studies show that its ancestors were probably aerobic bacteria. The evolution of C. botulinum towards an anaerobic form can be explained by the progressive adaptation to oxygen-deprived environments, where the bacterium has found a selective advantage by exploiting specific ecological niches, such as soils and deep sediments.

Growth and Sporulation

Clostridium botulinum grows under optimal anaerobic conditions, with an optimum growth temperature of between 35°C and 40°C for mesophilic strains. In an anaerobic environment, the bacteria use organic compounds such as pyruvic acid to produce energy through fermentation. Growth is also influenced by the pH of the medium, which should be between 4.6 and 8.5.

When confronted with unfavourable environmental conditions, such as the presence of oxygen, extreme temperatures or a lack of nutrients, C. botulinum enters a sporulation phase. Sporulation is a complex process whereby the bacterium forms resistant endospores capable of surviving hostile environmental conditions. These endospores can remain viable for years and germinate when conditions become favourable again. This ability to sporulate ensures the bacterium's persistence and dissemination in a variety of environments.

Habitat and Environmental Factors

Clostridium botulinum is ubiquitous, occurring mainly in soils, aquatic sediments and environments rich in organic matter. The bacterium thrives in natural anaerobic niches such as the seabed, marshes and animal intestines.

Environmental factors strongly influence the survival and growth of C. botulinum. In the absence of oxygen, the bacterium can proliferate in conditions of low acidity, moderate temperatures and the presence of organic matter. Temperature plays a crucial role: some psychrotrophic strains can develop at temperatures as low as 3°C, while others, which are thermophilic, prefer higher temperatures, up to 45°C.

Historically, the ancestors of Clostridium botulinum were probably aerobic and evolved towards an anaerobic form in response to environmental pressures. The ability of these bacteria to exploit oxygen-deprived niches enabled them to adapt and survive in a variety of habitats. This evolutionary transition to anaerobiosis was facilitated by the bacterium's ability to form spores, ensuring its persistence despite unfavourable conditions.

Potential presence in the oral cavity

The oral cavity, although for the most part aerobic, can contain anaerobic microenvironments, particularly in periodontal pockets and deep caries. These conditions are generally conducive to the growth of specific anaerobic bacteria such as Streptococcus mutans and Lactobacillus, which are responsible for dental caries by producing acids that demineralise tooth enamel.

Although Clostridium botulinum is rarely associated with dental infections, its theoretical presence in the oral cavity cannot be totally ruled out. Strict anaerobic conditions, combined with decomposing organic matter, could provide a favourable environment for its survival. However, typical dental infections do not provide ideal conditions for the development of C. botulinum.

Modern dental care, including the treatment of caries and the removal of necrotic tissue, considerably reduces the risk of colonisation by atypical pathogens such as C. botulinum. In cases of severe infection, antibiotics such as penicillins and metronidazoles are often used to target the anaerobic bacteria involved in dental infections.

This section explores the possibility, albeit rare, of Clostridium botulinum being present in the oral cavity and highlights the importance of understanding the specific environmental conditions required for its survival and proliferation.

Chapter 2: Mechanism of Botulinum Toxin

Structure and Types of Botulinum Toxins (A-G)

Botulinum toxins are a family of neurotoxins produced by the bacterium Clostridium botulinum. There are seven distinct types of botulinum toxin, designated A to G. Each of these toxins has a similar protein structure but specific variations that give them unique properties and mechanisms of action.

Botulinum toxin consists of two polypeptide chains: a heavy chain (100 kDa) and a light chain (50 kDa), linked by a disulphide bond. The heavy chain is responsible for binding and translocating the toxin within nerve cells, while the light chain has a protease activity that cleaves proteins essential for the release of the neurotransmitter acetylcholine.

In bioinorganic terms, it is interesting to note that botulinum toxins are metalloproteins. They contain zinc ions (Zn^{2+}) in their active site, which are essential for their enzymatic activity. Zinc plays a crucial role in stabilising the structure of the toxin and catalysing the cleavage of target proteins.

Mechanism of Action: How the Toxin Affects Nerve Cells

The mechanism of action of botulinum toxin is based on a series of precise steps:

1   Binding: The toxin's heavy chain binds specifically to receptors on the nerve cell membrane at neuromuscular junctions.

2   Endocytosis: The toxin is internalised by endocytosis, forming a vesicle inside the nerve cell.

3   Translocation: The acidification of the vesicle causes the heavy chain to form a transmembrane channel, allowing the light chain to enter the cell's cytoplasm.

4   Protein cleavage: The light chain, containing a zinc ion in its active site, specifically cleaves SNARE (Soluble N-ethylmaleimide-sensitive factor Attachment protein REceptor) proteins such as SNAP-25, syntaxin or synaptobrevin. These proteins are crucial for the fusion of synaptic vesicles with the plasma membrane, a process required for the release of acetylcholine.

5   Inhibition of acetylcholine release: By cleaving SNARE proteins, the toxin prevents the release of acetylcholine, leading to flaccid paralysis of the affected muscles.

Differences between Types of Toxins and their Effects

Each type of botulinum toxin (A-G) differs slightly in its protein structure, its specificity for cleaving SNARE proteins and its affinity for neuronal receptors. These differences influence the efficacy, duration of action and clinical applications of each toxin.

-Toxin A: The most widely used clinically, it targets SNAP-25 and has a long duration of action, making it effective in the treatment of a variety of medical conditions, including movement disorders and facial wrinkles.

-Toxin B: Targets synaptobrevin and is used as an alternative when patients develop resistance to toxin A.

-C-G toxins: Less common in clinical practice, they have variations in their SNARE protein cleavage site and have more specific or experimental applications.

The bioinorganic differences between these toxins, in particular the composition of metal ions and the structure of the active sites, determine their target recognition mechanism and their enzymatic efficacy. For example, variations in the coordination of zinc ions within light chains directly influence the proteolytic activity of each toxin.

Chapter 3: Bioinorganics of Clostridium botulinum

Clostridium botulinum is a strict anaerobic bacterium that possesses several metal metalloproteins and enzymes essential for its metabolism and survival in oxygen-deprived environments. These bioinorganic components play a crucial role in various cellular processes, including the regulation of the response to oxidative stress and pathogenesis.

Metalloproteins and Metallic Enzymes

Ferritin and Iron Storage

Clostridium botulinum possesses ferritin, a protein capable of storing iron in the form of ferricity, in order to maintain iron homeostasis under anaerobic conditions. Iron is an essential cofactor for several enzymes and cellular processes, including the response to oxidative stress. Iron is involved in the reduction of peroxides, protecting bacteria against oxidative damage. Ferritin plays a crucial role in regulating intracellular iron levels, ensuring an adequate supply of iron for metabolic processes while avoiding the toxic effects of excess iron.

Metalloenzymes Involved in Energy Metabolism :

- **Hydrogenase**: This enzyme contains iron-sulphur and nickel-iron metal centres, involved in the metabolism of hydrogen, an essential energy source for the bacteria in the absence of oxygen. Clostridium botulinum uses hydrogenase to metabolise hydrogen by producing energy in the form of ATP, thus contributing to the bacterium's survival in anaerobic environments.

Nitrogenase: Another enzyme crucial to Clostridium botulinum is nitrogenase, which contains iron-molybdenum. Nitrogenase enables the bacteria to fix atmospheric nitrogen, an essential source of nitrogen for the synthesis of amino acids and other biological compounds essential for bacterial growth. This nitrogen fixation process is vital because it enables the bacteria to grow in environments where organic nitrogen sources are limited.

Metabolism of Sulphur and Metals

Clostridium botulinum also uses metalloenzymes to metabolise sulphur and other metals such as manganese and zinc, which are essential for various biochemical reactions and defence against environmental stresses. For example, enzymes such as manganese-containing superoxide dismutases (SODs) help neutralise free radicals produced during oxidative stress. In addition, zinc is often involved in the regulation of enzyme activity, particularly in enzymes involved in the synthesis of nucleic acids and proteins.

Together, these metalloproteins and metal enzymes demonstrate the remarkable adaptation of Clostridium botulinum to its strict anaerobic environment, using metals as cofactors for essential metabolic processes and protecting itself against environmental stresses. Understanding these bioinorganic mechanisms not only sheds light on the fundamental biology of the bacterium, but also opens up prospects for the future biotechnological exploitation of these processes. ..

Regulation and Adaptation to Oxidative Stress

Because of its anaerobic habitat, Clostridium botulinum has developed sophisticated biochemical strategies to protect itself against oxidative stress, including the use of metalloproteins such as catalases, superoxide dismutases and metal-based enzymes to neutralise reactive oxygen species.

Biotechnological implications

Understanding the role of metalloproteins and metal enzymes in Clostridium botulinum opens up promising avenues for metabolic engineering and biotechnology. For example, optimising the enzymes involved in metal metabolism could improve the production of biomass and compounds of industrial interest.

Conclusion

In summary, Clostridium botulinum is highly dependent on metallic metalloproteins and enzymes for its metabolism and survival in an anaerobic environment. These bioinorganic components play an essential role in various cellular processes, from the regulation of energy metabolism to protection against oxidative stress. Their in-depth study opens the way to new biotechnological applications and a better understanding of the biology of this pathogenic bacterium.

Chapter 4: Biochemical applications and Botulism

Clostridium botulinum and its toxin, botulinum toxin, have a variety of applications in different fields, mainly due to their unique properties of modulating neurotransmissions and muscle activity. Because of these characteristics, they are used to treat muscle spasms and wrinkles, as well as in other therapeutic and cosmetic contexts.

Botulinum toxin acts primarily by inhibiting the release of acetylcholine at neuromuscular junctions, resulting in a temporary reduction in muscle contractions. This mechanism is used to reduce excessive muscle activity in conditions such as localised muscle spasms and dystonia. For example, in situations of involuntary muscle contraction such as spasmodic torticollis, botulinum toxin is used to reduce the frequency and intensity of contractions.

In addition to its effects on muscle activity, botulinum toxin is also used to alleviate chronic pain conditions such as migraines. It is applied locally, usually by injection, precisely targeting the muscle fibres concerned without disrupting overall muscle function. The effects begin to appear a few days after the injection, reaching their peak after one to two weeks.

Studies have confirmed the effectiveness of botulinum toxin in a variety of applications, showing a marked improvement in symptoms and quality of

21

life for those treated. These results are influenced by factors such as the dose administered and the injection method, as well as individual response to treatment.

Clostridium botulinum is also known to be the source of a potentially fatal condition called botulism, which manifests itself in different ways:

Food Botulism: Caused by the ingestion of food containing botulinum toxin, often as a result of poor food preservation or preparation.

Wound botulism: Resulting from the penetration of spores into an open wound, where they produce toxin.

Infantile botulism: Rare form affecting infants, often linked to ingestion of spores in foods such as honey.

Botulism by inhalation: Caused by inhalation of the toxin, generally in a laboratory or industrial setting.

Symptoms of botulism include muscle weakness, blurred vision, difficulty swallowing, dry mouth and progressive paralysis. Diagnosis is often based on these signs, confirmed by detection of the toxin in body secretions or food.

Treatment of botulism involves the administration of a specific antitoxin to neutralise the circulating toxin. Intensive care is often required, including respiratory support and close medical supervision. Long-term rehabilitation may be required to restore muscle function.

In conclusion, Clostridium botulinum and its toxin have significant applications in various fields, but they also present serious risks of botulism. A thorough understanding of their biochemistry and pharmacology is essential to maximise their benefits while minimising the associated risks.

Chapter.5. Safety and regulations

Production and Purification of Botulinum Toxin

The production and purification of botulinum toxin represent significant challenges due to the toxin's potentially high toxicity and the risks associated with handling it.

Botulinum toxin is produced by strains of Clostridium botulinum under strictly controlled conditions. The bacteria produce the toxin in the form of an inactive precursor, which is then purified to remove potential impurities while retaining the toxin's biological activity.

Regulatory aspects and safety measures

Because of its toxicity, botulinum toxin production, transport and clinical and cosmetic use are strictly regulated. The regulatory authorities impose strict standards to ensure the safety of workers and the public.

Safety measures include the use of personal protective equipment (PPE) when handling the toxin, as well as strict procedures for production and purification. Production facilities must comply with high biosafety standards to prevent accidents and toxin leaks.

Risk Assessment and Management in Clinical and Cosmetic Use

The clinical and cosmetic use of botulinum toxin involves careful risk assessment and appropriate management to minimise adverse effects.

The main risks include diffusion of the toxin to non-target areas, causing excessive muscle weakness, as well as potential allergic reactions. Practitioners must be trained in the safe administration of the toxin and the management of any complications.

Problems in Cosmetics and the Use of Nitrates

Cosmetic products containing botulinum toxin should be used with caution due to the potential risk of diffusion and adverse effects, even at a distance from the injection site. Cosmetic regulations vary from country to country, but all require certification and appropriate training for the use of botulinum toxin.

Nitrates are often used to combat Clostridium botulinum in food products. However, there are risks associated with their use, as nitrates can be toxic in high concentrations. They can also react with proteins to form carcinogenic compounds such as nitrosamine.

In conclusion, safety and regulation around botulinum toxin are crucial to ensure its safe and effective use in clinical and cosmetic applications. The production and handling of the toxin must follow strict protocols to minimise risks to public health and the environment.

Chapter 6. Future prospects

New Applications Research

Research into botulinum toxin has opened up exciting new avenues for its use in a variety of medical and non-medical fields. Potential new applications include:

Treatment of neurological diseases: Exploratory studies are underway to assess the efficacy of botulinum toxin in the treatment of neurological diseases such as chronic migraine, dystonia and even treatment-resistant depression.

Dermatological applications: In dermatology, botulinum toxin is being studied for its anti-ageing effects and for the treatment of skin conditions such as hyperhidrosis (excessive sweating) and non-surgical facelifts.

-Specific therapies: Research is underway to develop targeted therapies, such as the administration of genetically modified toxins to extend their duration of action and improve their efficacy.

Technological advances in Botulinum Toxin

Technological advances in the field of botulinum toxin are aimed at improving its production, stability and clinical use:

-Improved production: Recombinant production methods make it possible to obtain larger, purer quantities of botulinum toxin, reducing production costs and the risks associated with the use of pathogenic bacteria.

-Nanotechnology and administration: Advanced nanotechnology techniques are being studied to improve the administration of botulinum toxin, by increasing its bioavailability and specifically targeting the areas to be treated.

Non-invasive delivery technologies: Non-invasive delivery devices are being developed to facilitate the delivery of botulinum toxin, reducing the need for traditional injection techniques.

Ethical considerations and future challenges

Despite its many promising applications, the use of botulinum toxin raises ethical questions and challenges:

Safety and Risks: It is essential to continue studying the long-term effects of botulinum toxin, in particular its effects on the development of children and its long-term use in adults.

-Access and Equity: Access to botulinum toxin treatments may be limited due to their high cost, creating disparities in access to healthcare.

-Education and training: Proper training of healthcare professionals is necessary to ensure the safe and effective use of botulinum toxin, while minimising the risks to patients.

Conclusion

In conclusion, ongoing research into botulinum toxin and its technological advances are paving the way for potential new applications in various medical and aesthetic fields. However, it is crucial to address the ethical challenges, improve access to treatment and ensure the safe and ethical use of this powerful neurotoxin.

Conclusion

This book has explored in depth the many facets of Clostridium botulinum and its neurotoxin, botulinum toxin, through a lens of bioinorganic chemistry. From the biology and morphology of the bacterium to the structure and mechanism of action of the toxin, each chapter offered a detailed analysis supported by recent scientific advances.

We have discovered how this anaerobic bacterium has evolved to survive in hostile environments, using metals and metalloenzymes for its metabolism and protection against environmental stresses. We have explored in depth the mechanisms by which botulinum toxin disrupts nerve transmission, leading to innovative applications in the treatment of a variety of conditions.

Throughout this book, we have also examined the ethical and regulatory challenges associated with the use of botulinum toxin, as well as the technological advances that are transforming its use. Promising future prospects are emerging, with research into new applications and technological advances aimed at improving the safety and efficacy of this powerful neurotoxin.

Finally, this book has highlighted the crucial importance of ongoing research and education in the field of bioinorganic chemistry, underlining

32

the interdisciplinarity required to understand and exploit the potential of Clostridium botulinum and botulinum toxin.

As new discoveries emerge and technology advances, it is clear that Clostridium botulinum and its botulinum toxin will continue to play a central role in scientific research, offering new perspectives for treating disease and improving quality of life.

This book aims to serve as a comprehensive and informative guide for researchers and students interested in the fascinating role of bioinorganic chemistry in the study of Clostridium botulinum and botulinum toxin, in the hope of inspiring new ideas and discoveries in this exciting and constantly evolving field.

Glossary :

Acetylcholine: Neurotransmitter involved in the transmission of nerve signals to muscles.

Pyruvic acid: A key organic compound in cellular energy metabolism.

Non-invasive administration: method of administering drugs that does not involve perforation or penetration of the body's tissues.

Anaerobic: Organism that can live and develop in the absence of oxygen.

Anaerobic: Growth conditions without oxygen.

Antitoxin: Antibody used to neutralise a specific toxin.

Bacillus: Rod-shaped bacterium.

Bioavailability: Quantity of drug that reaches the bloodstream and is available for use by the body.

Bioinorganics: Branch of biochemical chemistry that studies the roles of metals in biological systems.

Biosafety: A set of measures designed to prevent biological risks to human and animal health, as well as to the environment.

Catalysis: Acceleration of the rate of a chemical reaction by a catalyst.

Certification: Process of issuing a document attesting to the conformity of a product or process to specific standards.

Iron-sulphur metal centres: Clusters of iron and sulphur in enzymes that catalyse redox reactions.

Cofactor: Non-protein molecule required by an enzyme for its catalytic function.

Cosmetics: Products and treatments used to improve physical appearance.

Dystonia: A neurological disorder characterised by involuntary and prolonged muscle contractions.

Cervical dystonia: Neurological disorder characterised by involuntary muscle contractions in the neck, leading to abnormal movements and uncomfortable postures.

Focal dystonia: Motor disorder characterised by prolonged, involuntary muscle contractions.

Endocytosis: Process by which a cell engulfs external particles by encircling them with its cell membrane.

Endospore: resistant structure formed inside certain bacteria, enabling them to survive in unfavourable conditions.

Anaerobic environments: Oxygen-deprived environments.

Health equity: Principle according to which all individuals should have equal access to health services, regardless of their socio-economic status or ethnic origin.

Fermentation: Anaerobic metabolic process in which micro-organisms convert organic compounds into energy.

Ferritin: Iron storage protein which maintains iron homeostasis in cells.

Nitrogen fixation: Conversion of atmospheric nitrogen into compounds that can be used by organisms.

Appropriate training: Adequate learning and practice of the procedures required to use a specific product or technique safely.

Vocational training: The process of acquiring the knowledge, skills and attitudes needed to practise a specific profession.

Gram-positive: Classification of bacteria that retain the crystal violet stain in Gram staining, due to the structure of their cell wall.

Hydrogenase: Enzyme using metal centres to catalyse the hydrogen reaction.

Hyperhidrosis: Medical condition characterised by excessive sweating.

Intramuscular injection: introduction of a drug directly into a muscle.

Food poisoning: Illness caused by the ingestion of food toxins.

Disulphide ion: Chemical bond containing two sulphur atoms linked together.

Ferric ion: Oxidised form of iron used for storing and transporting iron.

Zinc ion (Zn^2+): Metal ion crucial for the catalytic activity of botulinum toxins.

Acute respiratory illness: Severe illness affecting the lungs and breathing.

Mesophilic: Microorganism that prefers moderate growth conditions, typically between 20°C and 45°C.

Metabolism: All the chemical reactions that take place in an organism to maintain life.

Energy metabolism: All the chemical reactions that take place in cells to produce energy.

Metalloproteins: Proteins containing metal ions in their structure.

Metronidazole: Antibiotic and antiprotozoal used to treat various bacterial and parasitic infections.

Honeydew: Food produced by insects and containing a high percentage of sucrose.

Nanotechnology: Field of science and technology that manipulates matter on a nanometric scale to create new materials and devices.

Neurotoxin: Toxic substance that acts specifically on the nervous system.

Neurotransmitter: Chemical substance released by neurons to transmit a signal to other cells.

Ecological niche: Role and position of an organism in its environment, including its interactions with other organisms and abiotic factors.

Nickel-iron: Metal centre in enzymes, essential for hydrogen metabolism.

Nitrate: Chemical compound used as a food preservative, sometimes used to inhibit the growth of Clostridium botulinum in food products.

Nitrogenase: Enzyme containing iron-molybdenum, which catalyses the biological fixation of atmospheric nitrogen.

Nitrosamine: Carcinogenic chemical compound formed by the reaction between nitrates and amines.

Flaccid paralysis: A type of paralysis characterised by a loss of muscle tone and generalised weakness.

Penicillins: Group of antibiotics derived from Penicillium used to treat bacterial infections.

Peptidoglycan: Major component of the cell wall of Gram-positive bacteria, conferring rigidity and protection.

Protease: Enzyme which catalyses the cleavage of peptide bonds in proteins.

Protein: Biological macromolecule made up of one or more chains of amino acids.

Psychrotroph: Micro-organism capable of growing at low temperatures, close to 0°C.

Free radicals: Highly reactive chemical species containing one or more unpaired electrons.

Allergic reactions: exaggerated immune responses by the body to certain foreign substances.

Redox reaction: Chemical reaction in which electrons are transferred between the reactants.

Neuronal receptor: cell surface or intracellular protein which receives signals to provoke a cellular response.

Rehabilitation: The process of restoring normal function after illness or injury.

Public health risks: Potential health hazards for the general population.

Public health: Discipline which aims to improve the health of the population as a whole, with an emphasis on preventing disease and promoting well-being.

SNARE: Soluble N-ethylmaleimide-sensitive factor Attachment protein REceptor involved in synaptic vesicle fusion.

Sporulation: Process by which certain bacteria form resistant spores to survive in unfavourable environmental conditions.

Muscle spasms: Involuntary and often painful muscle contractions.

Strabismus: Eye disorder in which the eyes are not aligned correctly and look in different directions.

Oxidative stress: Cell damage caused by free radicals and other reactive oxygen species.

Superoxide dismutase (SOD): Enzyme which catalyses the dismutation of superoxide into oxygen and hydrogen peroxide.

Botulinum toxin: Neurotoxin produced by Clostridium botulinum, responsible for botulism and used for therapeutic and cosmetic purposes.

Genetically modified toxin: A toxin whose gene has been altered to improve its pharmacological properties or reduce its toxicity.

Targeted therapy: medical approach that specifically aims to treat a disease or disorder without affecting the surrounding healthy tissue.

Thermophile: Organism that prefers high temperatures for its growth.

Ubiquitous: An organism that can be found in many different types of environment.

Synaptic vesicle: Small vesicle containing neurotransmitters at the neuromuscular junction, ready to be released when the nerve signal is transmitted.

References :

1   Dover, N., Barash, J. R., Hill, K. K., Detter, J. C., & Arnon, S. S. (2009). Mechanisms of Botulinum Neurotoxin Action and Detection Methods. Frontiers in Bioscience, 14, 192-207. https://doi.org/10.2741/3262

2   Foster, K. A., & Fung, K. K. (Eds.). (2005). Botulinum Toxin: Therapeutic Clinical Practice and Science. CRC Press.

3   Nigam, P. K., & Nigam, A. (2010). Botulinum Toxin: A Clinical Overview. American Journal of Clinical Dermatology, 11(3), 149-159. https://doi.org/10.2165/11319440-000000000-00000

4   Van Ermengem, E. (1897). Über einen neuen anaëroben Bacillus und seine Beziehungen zum Botulismus. Zeitschrift für Hygiene und Infektionskrankheiten, 26(1), 1-56. https://doi.org/10.1007/BF02243160

5   Kerner, J. (1822). Neue Beobachtungen über die in Württemberg so häufig vorfallenden tödlichen Vergiftungen durch den Genuss geräucherter Würste. Tübingen.

6   Centers for Disease Control and Prevention. (2020). Botulism. Retrieved from https://www.cdc.gov/botulism/index.html

7   World Health Organization. (2019). Botulinum Toxin. Retrieved from https://www.who.int/biologicals/areas/toxins/botulinum/en/.

8.   Dover, N., Barash, J. R., Hill, K. K., Detter, J. C., & Arnon, S. S. (2009). Mechanisms of Botulinum Neurotoxin Action and Detection Methods. Frontiers in Bioscience, 14, 192-207. https://doi.org/10.2741/3262

9.   Foster, K. A., & Fung, K. K. (Eds.). (2005). Botulinum Toxin: Therapeutic Clinical Practice and Science. CRC Press.

10.   Nigam, P. K., & Nigam, A. (2010). Botulinum Toxin: A Clinical Overview. American Journal of Clinical Dermatology, 11(3), 149-159.

11.   Van Ermengem, E. (1897). Über einen neuen anaëroben Bacillus und seine Beziehungen zum Botulismus. Zeitschrift für Hygiene und Infektionskrankheiten, 26(1), 1-56. https://doi.org/10.1007/BF02243160

12.   Kerner, J. (1822). Neue Beobachtungen über die in Württemberg so häufig vorfallenden tödlichen Vergiftungen durch den Genuss geräucherter Würste. Tübingen.

13.   Centers for Disease Control and Prevention. (2020). Botulism. Retrieved from https://www.cdc.gov/botulism/index.html

14.   World Health Organization. (2019). Botulinum Toxin. Retrieved from https://www.who.int/biologicals/areas/toxins/botulinum/en/

15.   Dover, N., Barash, J. R., Hill, K. K., Detter, J. C., & Arnon, S. S. (2009). Mechanisms of Botulinum Neurotoxin Action and Detection Methods. Frontiers in Bioscience, 14, 192-207.

16.   Foster, K. A., & Fung, K. K. (Eds.). (2005). Botulinum Toxin: Therapeutic Clinical Practice and Science. CRC Press.

17.   Nigam, P. K., & Nigam, A. (2010). Botulinum Toxin: A Clinical Overview. American Journal of Clinical Dermatology, 11(3), 149-159.

18.   Van Ermengem, E. (1897). Über einen neuen anaëroben Bacillus und seine Beziehungen zum Botulismus. Zeitschrift für Hygiene und Infektionskrankheiten, 26(1), 1-56. https://doi.org/10.1007/BF02243160

19.   Kerner, J. (1822). Neue Beobachtungen über die in Württemberg so häufig vorfallenden tödlichen Vergiftungen durch den Genuss geräucherter Würste. Tübingen.

20.   Centers for Disease Control and Prevention. (2020). Botulism. Retrieved from https://www.cdc.gov/botulism/index.html

21.   World Health Organization. (2019). Botulinum Toxin. Retrieved from https://www.who.int/biologicals/areas/toxins/botulinum/en/

22.   J. E. Tudor, J. A. Sherry (2004). Iron storage by Clostridium botulinum and its relation to spore germination. Anaerobe, 10(5), 317-324. https://doi.org/10.1016/j.anaerobe.2004.06.001

23.   M. R. Adams, S. P. Stacey (1995). Iron availability: a novel mechanism for control of Clostridium botulinum toxin production. Applied and Environmental Microbiology, 61(4), 1271-1273. https://doi.org/10.1128/AEM.61.4.1271-1273.1995

24.   K. Nakamoto (2009). Metalloenzymes: Bioinorganic Chemistry. In Inorganic Chemistry: A Textbook (5th ed., pp. 515-520). John Wiley & Sons.

25.   J. E. Tudor, J. A. Sherry (2004). Iron storage by Clostridium botulinum and its relation to spore germination. Anaerobe, 10(5), 317-324. https://doi.org/10.1016/j.anaerobe.2004.06.001

26.   M. R. Adams, S. P. Stacey (1995). Iron availability: a novel mechanism for control of Clostridium botulinum toxin production. Applied and Environmental Microbiology, 61(4), 1271-1273. https://doi.org/10.1128/AEM.61.4.1271-1273.1995

27.   K. Nakamoto (2009). Metalloenzymes: Bioinorganic Chemistry. In Inorganic Chemistry: A Textbook (5th ed., pp. 515-520). John Wiley & Sons

28.   Dressler, D. (2020). Clinical applications of botulinum toxin. Current Opinion in Neurology, 33(4), 492-497.

29.   Sobel, J. (2005). Botulism. Clinical Infectious Diseases, 41(8), 1167-1173. https://doi.org/10.1086/444507

30.   Pirazzini, M., Rossetto, O., & Eleopra, R. (2017). Botulinum neurotoxins: Biology, pharmacology, and toxicology. Pharmacological Reviews, 69(2), 200-235.

31. Dressler, D. (2020). Clinical applications of botulinum toxin. Current Opinion in Neurology, 33(4), 492-497. https://doi.org/10.1097/WCO.0000000000000821

32. Sobel, J. (2005). Botulism. Clinical Infectious Diseases, 41(8), 1167-1173.

33. Pirazzini, M., Rossetto, O., & Eleopra, R. (2017). Botulinum neurotoxins: Biology, pharmacology, and toxicology. Pharmacological Reviews, 69(2), 200-235.

34. Dressler, D. (2020). Clinical applications of botulinum toxin. Current Opinion in Neurology, 33(4), 492-497. https://doi.org/10.1097/WCO.0000000000000821

35. Pirazzini, M., Rossetto, O., & Eleopra, R. (2017). Botulinum neurotoxins: Biology, pharmacology, and toxicology. Pharmacological Reviews, 69(2), 200-235.

Appendix

1.　　Books :

-Botulinum Toxin: Therapeutic Clinical Practice and Science by Keith A. Foster and Keith K. Fung This book provides a detailed exploration of the clinical applications of botulinum toxin, as well as the science behind its therapeutic use.

-Handbook of Clostridia by P. Durre This handbook covers a comprehensive range of information on clostridia, including Clostridium botulinum, with particular emphasis on biology, pathogenesis and industrial applications.

2.　　research articles :

-Mechanisms of Botulinum Neurotoxin Action and Detection Methods" by Dover N, Barash JR, Hill KK, Detter JC, Arnon SS. This article explores the mechanisms of botulinum neurotoxin action and associated detection methods, providing an in-depth analysis of the neurobiological pathways affected.

-Botulinum Toxin: A Clinical Overview" by Nigam PK, Nigam A. This article provides a comprehensive clinical overview of botulinum toxin, covering its therapeutic applications in various medical fields.

3　　Websites :

Centers for Disease Control and Prevention (CDC) on botulism The CDC provides up-to-date resources and guidelines on botulism, including the latest information on the epidemiology, diagnosis and treatment of this disease.

-World Health Organization (WHO) guidelines on botulinum toxin The WHO guidelines provide global recommendations on the safe and effective use of botulinum toxin and the management of botulism.

Printed by Books on Demand GmbH, Norderstedt / Germany